MÁRQUEZ Henri Matisse WILLIAM CARLOS WILLIAMS Josef Engelha

Dante Gabriel Rossetti PABLO NERUDA Luca Forte FEDERICO GAR

omenico Ghirlandaio ALICE B. TOKLAS William Hooker MARY FORS

DE Raphaelle Peale WILLIAM SHAKESPEARE D.Frankcom JAMES DE COQ

tiste Siméon Chardin WALLACE STEVENS Redoute Willie

AN TABLADA Jessica Hayllar SHARON OLDS SAP

slow Homer ERNEST HEMINGWAY Albrecht ILLIAM SHAKESPE

MÁRQUEZ Henri Matisse WILLIAM CARLOS WILLIAMS Josef Engelha

Dante Gabriel Rossetti PABLO NERUDA Luca Forte FEDERICO GAR

omenico Ghirlandaio ALICE B. TOKLAS William Hooker MARY FORS

DE Raphaelle Peale WILLIAM SHAKESPEARE D.Frankcom JAMES DE COQ

tiste Siméon Chardin WALLACE STEVENS Redoute JOHN MILTON Willie

AN TABLADA Jessica Hayllar SHARON OLDS William B. Hough SAP

slow Homer ERNEST HEMINGWAY Albrecht Dürer WILLIAM SHAKESPE

MÁRQUEZ Henri Matisse WILLIAM CARLOS WILLIAMS Josef Engelha

Dante Gabriel Rossetti PABLO NERUDA Luca Forte FEDERICO GAR

THE Frans Snyders WILLIAM BLAKE Henri Matisse PABLO NERU

nslow Homer ERNEST HEMINGWAY Albrecht Dürer WILLIAM SHAKESPEA

ELIOT Dorothy Stevens COUNTESS VON ARNIM Frida Kahlo GABRIEL G

QUEZ Henri Matisse WILLIAM CARLOS WILLIAMS Josef Engelha

nando Botero ERNEST HEMINGWAY Paul Gauguin GEORGE ROBERT GISSI

nte Gabriel Rossetti PABLO NERUDA Luca Forte FEDERICO GARCIA LOR

nslow Homer DELLA LUTES Albrecht Dürer AGNES MAXWELL-HA

menico Ghirlandaio ALICE B TOKLAS William Hooker MARY FORS

re Leighton SAMUEL REYNOLDS HOLE Eloise Harriet Stannard AND

Raphaelle Peale WILLIAM SHAKESPEARE L. Frankcom JAMES DE COQU

zabeth Rice SANDRA CISNEROS Vincenzo Campi JOHN DE MERS Je

ptiste Simeon Chardin WALLACE STEVENS Redoute JOHN MILT

lliam Hammer CLAUDE MCKAY A. Thouet JACOB CATS Giovan

rzoni JOSE JUAN TABLADA Jessica Hayllar SHARON OLDS William

ugh SAPPHO Frans Snyders WILLIAM BLAKE Henri Matisse GOET

O NERUDA Winslow Homer ERNEST HEMINGWAY Albrecht Dürer WILLI

THE CULTIVATED

GARDENER

Fruits

Edited by Kristin Joyce

A SWANS ISLAND BOOK

CollinsPublishersSanFrancisco

A Division of HarperCollins*Publishers*

Copyright ©1996 Kristin Joyce

Published in 1996 by
Collins Publishers San Francisco
1160 Battery Street
San Francisco, California 94111

Produced by Swans Island Books, Inc.
Belvedere, California 94920

Book design by Madeleine Corson Design
with special gratitude to Madeleine and Ann.

Additional thanks to:
Laurie Platt Winfrey and Robin Sand of Carousel Research, Inc.
Shellei Addison of Flying Fish Books
Julie Nathan of Swans Island Books

Library of Congress Cataloging-in-Publication Data
Fruits/ edited by Kristin Joyce
p. cm. – (The Cultivated gardener)
Swan's Island books
ISBN 0-00-225055-1
1. Fruits – Quotations, maxims, etc.
I. Joyce, Kristin. II. Series
PN6084. F83F78 1996
582. 13' 04166 – dc20   95-16290
Printed in Italy
1 3 5 7 9 10 8 6 4 2

"The spirits of the air live on the smells of fruit." Blake's vivid observation must surely stem from understanding that humankind also lives on, in fact *devours*, fruits. The spirits may imbibe the lush aromas, but we indulge in a total, sensory feast. We touch, hear, see and taste the intoxicating scents in the velvety feel of skin peeled from a sugar peach, the evocative crush of an opened watermelon, the suggestive shape of a comice pear, and the luscious taste of a mango's soft pulp. Drenched by fragrant juice, pungent seeds, and sweet meat, the *fruits* of this anthology evoke pleasure, both pure and bittersweet. ⌐ In thought, word and picture, most fruits are metaphors for desire. Rossetti's dark beauty, as a prime example, grasps an open pomegranate with seeds as red as her lips, adding a distinct sensuality to the love sonnet of Pablo Neruda: "I want to look back and see you in the branches. Little by little you turned into fruit…and my mouth will fill with the taste of you." We also read Shakespeare's suggestive verse, "So we grew together, like a double Cherry,…moulded on one stem," Sandra Cisneros' affectionate words, "If peaches had arms surely they would hold one another in their peach sleep," and Gabriel García Márquez's dramatic prose, "It was inevitable: the scent of bitter almonds always reminded him of the fate of unrequited love." ⌐ Celebration of fruit and the female form is a secondary theme, as might be expected, and for the artist, an obvious source of ripe delight. The shapes, colors and textures which describe peaches, berries, and pears, among others, often parallel the sensual descriptions of women. ⌐ There is possibly no greater statement than Botero's painting of the immense "Colombiana," her sumptuous, if not overripe, curves spanning from edge to edge, the tiniest green apple in hand. Vicenzo's buxom fruit vendor paired with James de Coquet's observation emphasizes the erotic undertones stating "Renoir used to say to young artists longing to paint — like him — the pinkish-brown tones of an opulent breast: 'First paint apples and peaches in a fruit bowl.'" ⌐ *The Cultivated Gardener: Fruits,* is one in a series of four books which includes single subject anthologies on *Flowers, Vegetables* and *Trees.* Each volume celebrates the gardens, both wild and tame, which cross every field of human endeavor and unite us with the world of nature. It is a sadness to be limited by ninety-six pages. From the pagan to the exotic to the vividly real, every morsel of text and art is worth savoring. May you relish and devour, for "It is no use making two bites of a cherry."

⌐ *K.J.*

The spirits of the air live on the smells of fruit...

I want to look back and see you in the branches.
Little by little you turned into fruit.
It was easy for you to rise from the roots,
singing your syllable of sap.

      Here you will be a fragrant flower first,
      changed to the statuesque form of a kiss,
      till the sun and the earth, blood and the sky, fulfill
      their promises of sweetness and pleasure, in you.

            There in the branches I will recognize your hair,
            your image ripening in the leaves,
            bringing the petals nearer my thirst,

                  and my mouth will fill with the taste of you,
                  the kiss that rose from the earth
                  with your blood, the blood of a lover's fruit.

                        PABLO NERUDA

"Doesn't the scent do one good?" went on his sister. "When I come into the garden on a morning like this, I have a feeling – *oh!* I can't describe it to you – perhaps you wouldn't understand –"

"It's as if nature were calling out to me,
like a friend, to come and enjoy what she has done.
I feel grateful for the things that earth offers me."

GEORGE ROBERT GISSING

Does not
the very fruit,
untainted by the arts
produce an impressive
sight for our being?...

JACOB CATS

I have eaten
the plums
that were in
the icebox

and which
you were probably
saving
for breakfast

Forgive me
they were delicious
s o   s w e e t
and so cold

WILLIAM CARLOS WILLIAMS

Bananas ripe and green, and ginger-root,
Cocoa in pods and alligator pears,
And tangerines and mangoes and grapefruit,
Fit for the highest prize at parish fairs,...

My eyes grew dim, and I could no more gaze;
A w a v e o f l o n g i n g through my body swept,
          And, hungry for the old, familiar ways,
                    I turned aside and bowed my head and wept.

CLAUDE MCKAY

Summer's loud laugh
Of scarlet ice
A melon
slice

JOSÉ JUAN TABLADA

Honey, pepper, leaf-green limes,
Pagan fruit whose names are rhymes,
Mangoes, breadfruit, ginger-roots,
Granadillas, bamboo-shoots,
      Cho-cho, ackees, tangerines,
      Lemons, purple Congo-beans,
      Sugar, okras, kola-nuts,
      Citrons, hairy coconuts,
            Fish, tobacco, native hats,
            Gold bananas, woven mats,
            Plantains, wild-thyme, pallid leeks,
            Pigeons with their scarlet beaks,
                  Oranges and saffron yams,
                  Baskets, ruby guava jams,
                  Turtles, goat-skins, cinnamon,
                  Allspice, conch-shells, golden rum.
                  Black skins, babel — *and the sun*
                  That burns all colours into one.

AGNES MAXWELL-HALL

A little sour is the juice of the pomegranate
like the juice of unripe raspberries.
                Waxlike is the flower
                Coloured as the fruit is coloured.
                        Close-guarded this item of treasure, beehive partitioned,
                        Richness of savour,
                        Architecture of pentagons.
                                The rind splits: out tumble the seeds,
                                In cups of azure some seeds are blood;
                                On plates of enamelled bronze,
                                        others are drops of gold.

ANDRÉ GIDE

It is said that
the only proper place
to eat a mango is
in the bathtub.

JOHN DE MERS

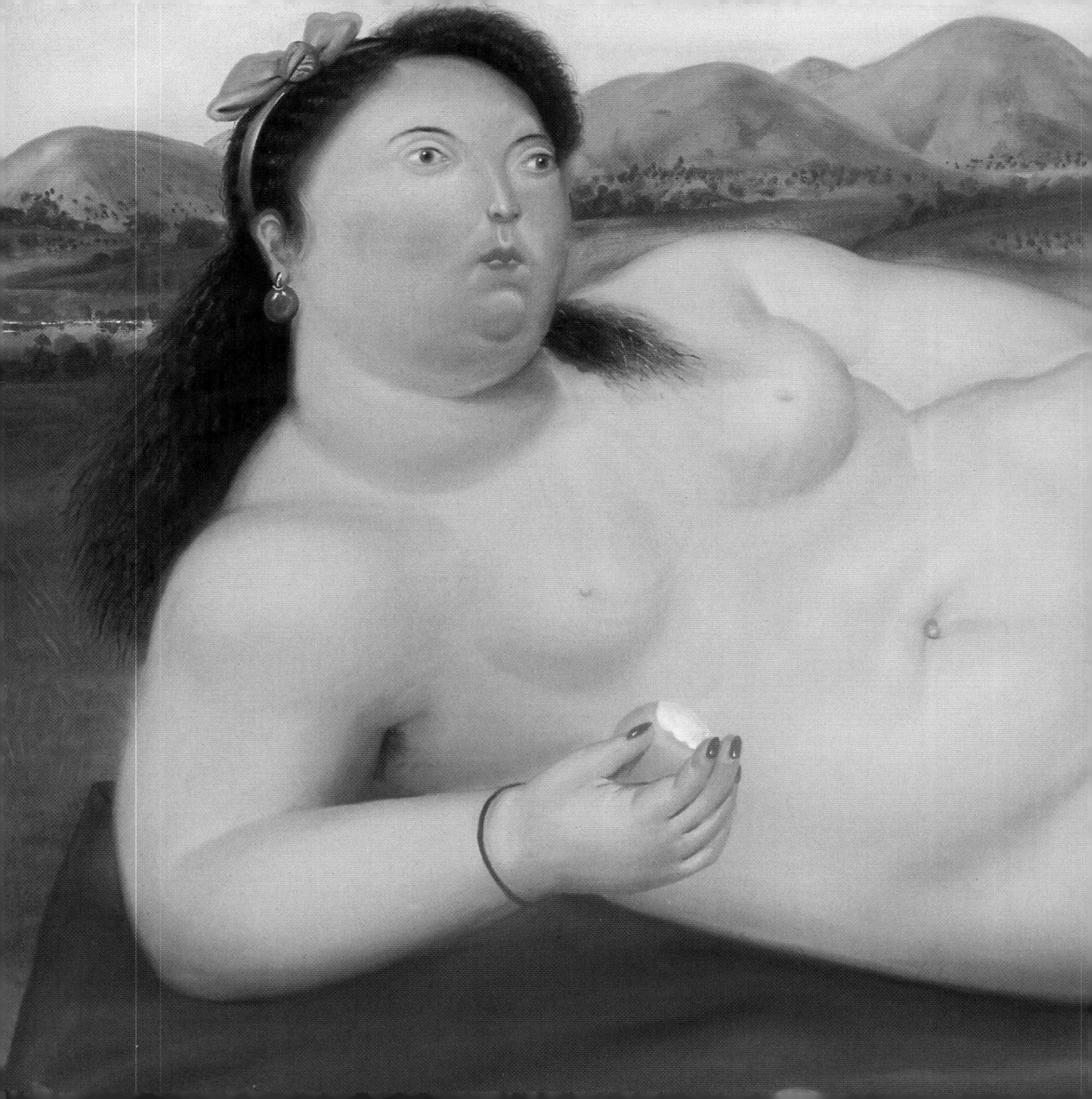

I Opusculum paedagogum.
The pears are not viols,
Nudes or bottles,
They resemble nothing else.

II They are yellow forms
Composed of curves
Bulging toward the base.
They are touched red.

III They are not flat surfaces
Having curved outlines.
They are round
Tapering toward the top.

IV In the way they are modelled
There are bits of blue.
A hard dry leaf hangs
From the stem.

V The yellow glistens.
It glistens with various yellows,
Citrons, oranges, and greens
Flowering over the skin.

VI The shadows of the pears
Are blobs on the green cloth
The pears are not seen
As the observer wills.

WALLACE STEVENS

Her rash hand in evil hour
Forth reaching to the fruit,
s h e   p l u c k e d ,   s h e   e a t ;
Earth felt the wound,
and Nature from her seat,
Sighing through all her works,
gave signs of woe
T h a t   a l l   w a s   l o s t .

JOHN MILTON

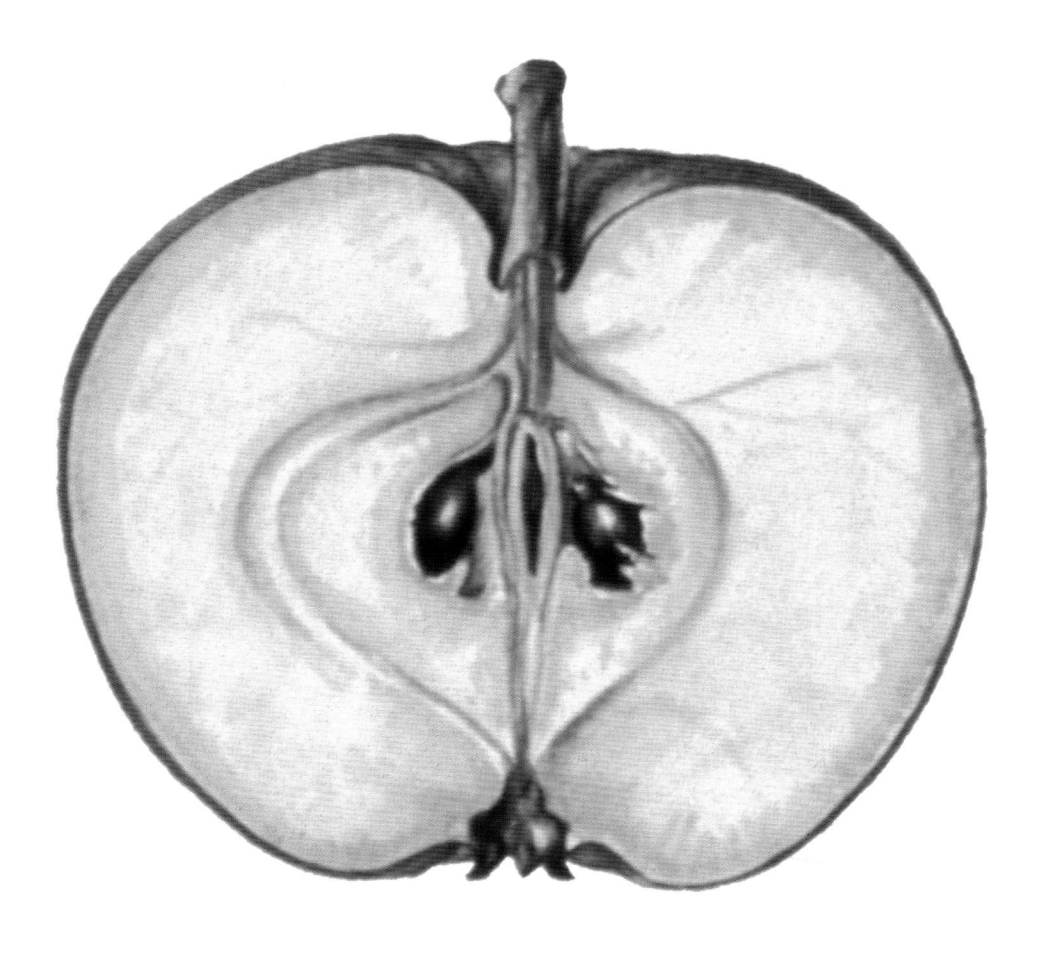

It is not graceful,
and it makes one hot;
but it is a blessed sort of work,
and if Eve had had a spade in Paradise
and known what to do with it,
we should not have had all
that sad business
of the apple.

COUNTESS VON ARNIM

Leave Crete,
Aphrodite,
and come to this
s a c r e d p l a c e
encircled by apple trees,
fragrant with offered smoke.

Here, cold springs
sing softly
amid the branches;
the ground is shady with roses;
from trembling young leaves
a deep drowsiness pours.

In the meadow,
horses are cropping the wildflowers of spring,
s c e n t e d f e n n e l
blows on the breeze.

In this place,
Lady of Cyprus, pour
the nectar that honors you
into our cups,
gold, and raised for the drinking.

SAPPHO

...like one of the best rooms in the finest museum
except there was a big fireplace and it was warm and comfortable
and they gave you good things to eat
and tea and natural distilled liqueurs made from
p u r p l e   p l u m s ,   y e l l o w   p l u m s
or wild raspberries.

These were fragrant, colorless alcohols
served from cut-glass carafes in small glasses
and whether they were
quetsche, mirabelle or framboise
they all tasted like the fruits they came from...

ERNEST HEMINGWAY

Whether the knife falls on the melon
or the melon on the knife,
the melon suffers.

AFRICAN PROVERB

Perhaps the only less than tasteful blow
to the raspberry's reputation was struck in the
political arena, where things so often turn sour.
In the 1840 presidential campaign in America, the Whig party
attacked opposing candidate Martin Van Buren for
overindulging in raspberries.

MARY FORSELL

I grow old... I grow old...
I shall wear the bottoms
of my trousers rolled.
Shall I part my hair behind?
Do I dare to eat a peach?

T.S. ELIOT

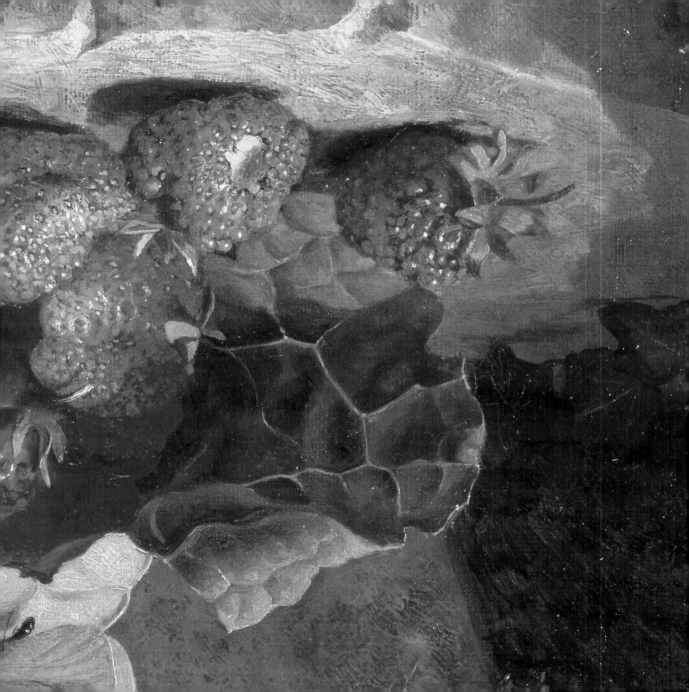

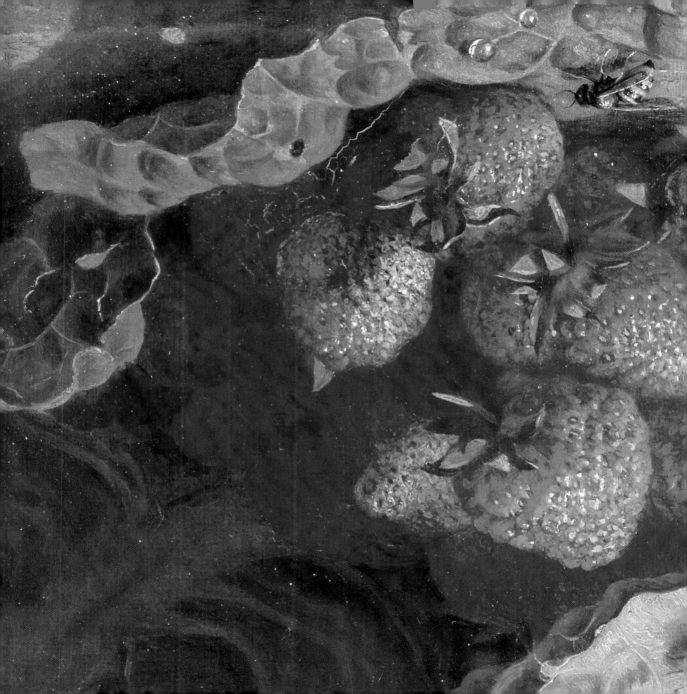

I asked a schoolboy, in the sweet summertide,
'what he thought a garden was for?' and he said, Strawberries.
His younger sister suggested Croquet and the elder Garden-parties.
The brother from Oxford made a prompt declaration in favour of Lawn Tennis
and Cigarettes, but he was rebuked by a solemn senior, who wore spectacles,
and more back hair than is usual with males, and was told that 'a garden was
designed for botanical research, and for the classification of plants.'
He was about to demonstrate the differences between the Acoty-
and the Monocoty-ledonous divisions when the collegian
remembered an engagement elsewhere.

I repeated my question to a middle-aged nymph,
who wore a feathered hat of noble proportions over a loose green tunic with a silver belt,
and she replied, with a rapturous disdain of the ignorance which presumed to ask —
'What is a garden for? For the soul, sir, for the soul of the poet! For visions of
the invisible, for grasping the intangible, for hearing the inaudible, for
exaltations above the miserable dullness of common life into the splendid regions
of imaginations and romance.'...

A capacious gentleman informed me that nothing in horticulture
touched him so sensibly as green peas and new potatoes,
and he spoke with so much cheerful candour that I could not
get angry; but my indignation was roused by a morose millionaire,
when he declared that of all his expenses he grudged most
the outlay of his confounded garden.

SAMUEL REYNOLDS HOLE

Give you a reason on compulsion! —
if reasons were as plentiful as Blackberries,
I would give no man a reason
upon compulsion.

WILLIAM SHAKESPEARE

The first rhubarb of the season
is to the digestive tract of the winter-logged *inner* man
what a good hot bath with plenty
of healing soap is to the *outer*
after a bout with a plough and harrow.
Even the tongue and teeth have a scrubbed feeling
after a dish of early rhubarb.

DELLA LUTES

If peaches had arms
surely they would hold one another
in their peach sleep.
                    And if peaches had feet
                    it is sure they would
                    nudge one another
                    with their soft peachy feet.
                                        And if peaches could
                                        they would s l e e p
                                        with their dimpled head
                                        on the other's
                                        each to each.
                                                        Like you and me.
                                                        And sleep and s l e e p.

                                                        SANDRA CISNEROS

Do you know that country where the lemon-trees flower,
And oranges of gold glow in the dark leaves,
And a gentle breeze blows
from blue heaven.

GOETHE

It was inevitable:
the scent of bitter almonds
always reminded him of the
fate of unrequited love.

GABRIEL GARCÍA MÁRQUEZ

The rind lies on the table where Liddy has left it
torn into pieces the size of petals and curved like petals,
rayed out like a full-blown rose, one touch will make it c o m e   a p a r t.
      The lining of the rind is wet and chalky as Devonshire cream,
      rich as the glaucous lining of a boiled egg,
      all that protein cupped in the ripped shell.
            And the navel, *torn out carefully*, lies there like a
            fat gold bouquet, and the scar of the stem,
            picked out with her nails, and still attached to the
            white thorn of the central integument, lies on the careful heap,
            a tool laid down at the end of a ceremony.

SHARON OLDS

The work in the vegetables —
Gertrude Stein was undertaking the care
of the flowers and box hedges — was a full time job and more.
Later it became a joke, Gertrude Stein asking me
what I saw when I closed my eyes, and I answered, Weeds.
That, she said was not the answer,
and so the weeds were changed to strawberries,
called by the French 'wood strawberries,'
are not wild but cultivated.
It took me an hour to gather a
small basket for Gertrude Stein's breakfast...
our young guests were told that if they cared to
eat them they should do the picking themselves.

ALICE B. TOKLAS

The spirits of the air live on the smells
of fruit; and Joy, with pinions light, roves round
The gardens, or sits s i n g i n g i n t h e t r e e s.
                              Thus sang the jolly Autumn as he sat;
                              Then rose, girded himself, and o'er the bleak
                              Hills fled from our sight; but left his g o l d e n l o a d.

                                                          WILLIAM BLAKE

August.
The opposing
of peach and sugar,
and the sun inside
the afternoon like
the stone in the fruit.

FEDERICO GARCÍA LORCA

Renoir used to say to young artists longing to paint
— like him — the pinkish-brown tones of an opulent breast:
"First paint apples and peaches in a fruit bowl."

JAMES DE COQUET

So we grew together,
        Like to a double Cherry, s e e m i n g   p a r t e d,
                But yet a union in partition;
                        Two lovely berries moulded on one stem.

WILLIAM SHAKESPEARE

It is no use making two bites of a cherry.

Painting Credits

•

Text Credits

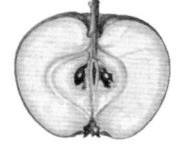

PAGE 66
Josef Engelhard, Dinner with Friends, 1903, Historisches Museum der Stadt, Vienna, BAL

PAGE 68
Clare Leighton, Picking Strawberries, 20th century, Courtesy of the Clare Leighton Estate

PAGE 70
Domenico Ghirlandaio, Birth of St. John, detail, 15th century, S. Maria Novella, Florence, AR/S

PAGES 72-73
Jean Baptiste Siméon Chardin, Basket of Peaches and Nuts, 1768, Musée du Louvre, Paris, AR/L

PAGE 75
Raphaelle Peale, Bowl of Peaches, Ackermann and Johnson, Ltd., London, BAL

PAGE 76
Vincenzo Campi, The Fruit Vendor, 16th century, Brera Gallery, Milan, AR/S

PAGES 80-81
Bartolomeo Bimbi, Cherries, Pitti Gallerie, Florence, AR/S

FOLLOWING CREDITS PAGE
Jan van Huysum, Flowers and Fruits, 18th century, Musée Fabre, Montpellier, AR/G

KEY TO ABBREVIATIONS

AR: Art Resource

AR/G: Art Resource/Giraudon

AR/L: Art Resource/Lessing

AR/M: Art Resource/Marburg

AR/NMAA: Art Resource/National Museum of American Art, Washington, D.C.

AR/S: Art Resource/Scala

AR/V&A: Art Resource/Victoria & Albert Museum, London

AKG: Archiv für Kunst und Geschichte

ARS: Artists Rights Society, NY

BAL: Bridgeman Art Library

Text Credits

PAGE 3

Pablo Neruda, *100 Love Sonnets*

PAGES 6-7

George Robert Gissing, *A Celebration of Gardens*

PAGE 9

Jacob Cats, *Gardening Through the Ages*

PAGE 11

William Carlos Williams, *The Collected Earlier Poems of William Carlos Williams*

PAGES 14-15

Claude McKay, *Art & Nature*

PAGE 19

José Juan Tablada, *Jane Grigson's Fruit Book*

PAGE 21

Agnes Maxwell-Hall, *3000 Years of Black Poetry*

PAGE 25

André Gide, *Jane Grigson's Fruit Book*

PAGE 26

John De Mers, *Dictionary of Famous Quotations*

PAGES 30-31

Wallace Stevens, *The Collected Poems of Wallace Stevens*

PAGE 33

John Milton, *Paradise Lost*

PAGE 37

Countess von Arnim, *Elizabeth and her German Garden*

PAGE 39

Sappho, *Into the Garden, A Wedding Anthology*

PAGES 42-43

Ernest Hemingway, *Gourmet*, April, 1992

PAGE 45

African Proverb, *Quotations of Wit and Wisdom: Know or Listen to Those Who Know*

PAGE 47

Mary Forsell, *Berries*

PAGE 48

T.S. Eliot, "*The Love Song of J. Alfred Prufrock*"

PAGES 52-53

Samuel Reynolds Hole, *Our Gardens*

PAGE 55

William Shakespeare, *Henry IV*

PAGE 57

Della Lutes, *The Country Kitchen*

PAGE 59

Sandra Cisneros, *My Wicked Wicked Ways*

PAGE 61

Goethe, *Wilhelm Meister*

PAGE 64

Gabriel García Márquez, *Love in the Time of Cholera*

PAGE 67

Sharon Olds, *The Gold Cell*

PAGE 69

Alice B. Toklas, *The Alice B. Toklas Cookbook*

PAGE 71 & BACK COVER

William Blake, *Herbs Through the Seasons at Caprilands*

PAGE 74

Federico García Lorca, *Roots and Wings*

PAGE 77

James De Coquet, *Jane Grigson's Fruit Book*

PAGE 79

William Shakespeare, *Midsummer Night's Dream*

PAGE 82

Proverb, *Gardener's Latin*

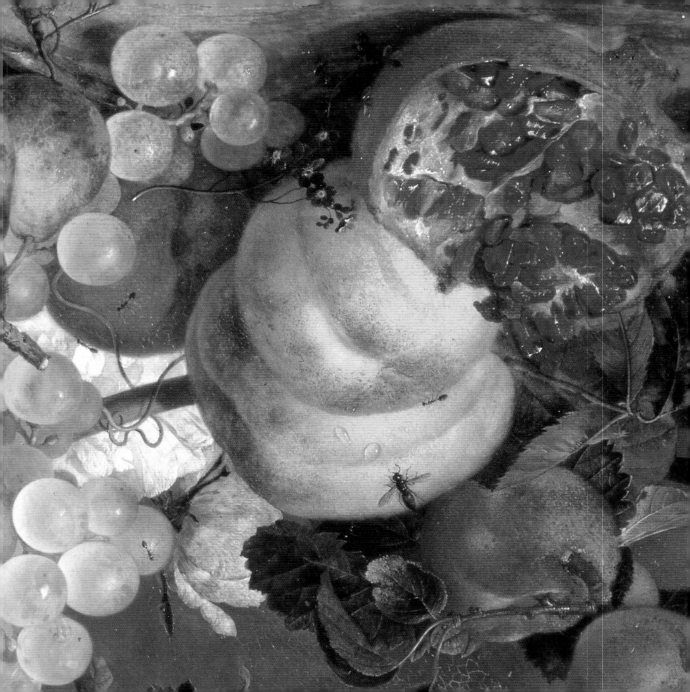

A S W A N S   I S L A N D   B O O K

Kristin Joyce is an author and book packager who produces
illustrated works for adults and children under her imprint Swans Island Books.
She has created and collaborated on over fourteen titles including this four volume
collection of select anthologies. *The Cultivated Gardener: Trees, Flowers, Fruits* and *Vegetables*
will be followed by a two-book sequel series, *The Cultivated Traveler,* in fall 1996.
Apart from books, Kristin relishes family life with her director-cinematographer
husband and their two wild and wonderful little ones. When time allows,
she swims, travels and cultivates two tiny knot gardens in Belvedere, California.

———————

Book designer Madeleine Corson has been creating
award-winning print work and packaging for over thirteen years.
She lives, works and dog walks in San Francisco.

ELIOT Dorothy Stevens COUNTESS VON ARNIM Frida Kahlo GABRIEL

rnando Botero ERNEST HEMINGWAY Paul Gauguin GEORGE ROBERT GI

rca Winslow Homer DELLA LUTES Albrecht Dürer AGNES MAXWELL-H

are Leighton SAMUEL REYNOLDS HOLE Eloise Harriet Stannard AND

lizabeth Rice SANDRA CISNEROS Vincenzo Campi JOHN DE MERS Jean

ammer CLAUDE MCKAY A. Theuet JACOB CATS Giovanna Garzoni JO

ETHE Frans Snyders WILLIAM BLAKE Henri Matisse PABLO NERUDA

ELIOT Dorothy Stevens COUNTESS VON ARNIM Frida Kahlo GABRIEL

rnando Botero ERNEST HEMINGWAY Paul Gauguin GEORGE ROBERT GI

rca Winslow Homer DELLA LUTES Albrecht Dürer AGNES MAXWELL-H

are Leighton SAMUEL REYNOLDS HOLE Eloise Harriet Stannard AND

lizabeth Rice SANDRA CISNEROS Vincenzo Campi JOHN DE MERS Jean

ammer CLAUDE MCKAY A. Theuet JACOB CATS Giovanna Garzoni JO

ETHE Frans Snyders WILLIAM BLAKE Henri Matisse PABLO NERUDA

ELIOT Dorothy Stevens COUNTESS VON ARNIM Frida Kahlo GABRIEL G

rnando Botero ERNEST HEMINGWAY Paul Gauguin GEORGE ROBERT